The Biosphere

Mary Kay Carson

Contents

A Living Planet

If you were traveling in a spaceship through Earth's neighborhood, you'd notice three similarly sized planets. But a closer look at Venus, Earth, and Mars would reveal very different worlds. While Venus is too hot and Mars too cold, Earth's temperature is just right—for life.

Earth is a world covered in life, thanks to its "just right" distance from the sun, plentiful water, breathable air, and nourishing land. The thin layer around Earth where life exists is called the **biosphere.** It includes all of Earth's life-forms as well as the water, air, rock, and soil they depend on.

But the biosphere is more than a collection of animals, plants, continents, and seas. It's a set of connected systems that recycle energy, water, even the stuff we are made of. The biosphere is a global **ecosystem,** the living part of our planet.

Venus's thick, smothering **atmosphere** makes it so hot that its oceans boiled away long ago.

Mars is too cold and its atmosphere too thin to support life on the surface.

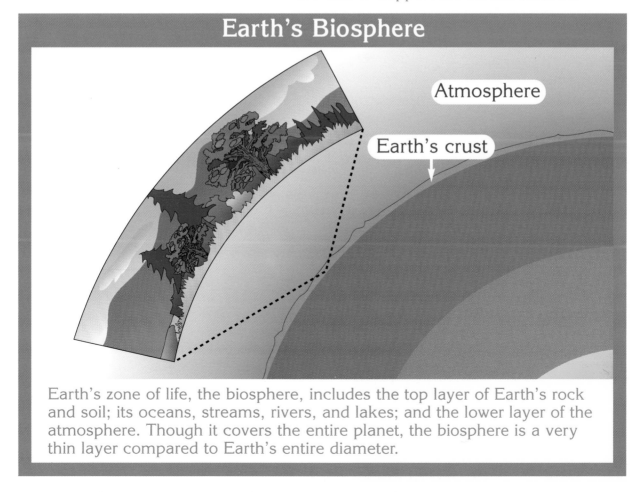

Earth's Biosphere

Atmosphere

Earth's crust

Earth's zone of life, the biosphere, includes the top layer of Earth's rock and soil; its oceans, streams, rivers, and lakes; and the lower layer of the atmosphere. Though it covers the entire planet, the biosphere is a very thin layer compared to Earth's entire diameter.

What Is Life?

The biosphere may be only a thin layer surrounding Earth, but it supports an amazing number of life-forms. Earth is home to millions of different species, or kinds, of plants, animals, fungi, bacteria, and other organisms. Scientists continue to find new species—some in places once thought to be too cold, too hot, too deep, or too high for life to exist.

Too hot to touch, the water in this Yellowstone National Park hot spring is teeming with life. The deep-blue color is the result of bacteria that thrive in water above 180°F. Around the pool's edges, where the temperatures are less hot, bacteria color the water green, yellow, orange, and brown.

The fangtooth lives in the frigid and dark deep sea. Its body is adapted to withstand the weight of hundreds of meters of heavy seawater above it.

It seems easy to recognize living from nonliving things, but defining life isn't so easy. Scientists consider something to be alive if it does most of the following: reproduces, grows, uses energy (sunlight or food), reacts to its environment, moves (toward food or sunlight), can regulate itself independently of its environment, and lives in an environment suitable to it.

Living things tend to regulate their bodies. Mammals and other warm-blooded animals maintain a steady body temperature that doesn't change, regardless of the outside temperature.

This crystal is in an environment that helps it grow, but it isn't alive. Why not?

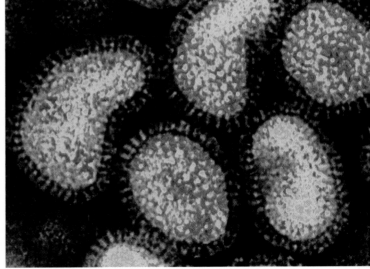

Viruses are tiny bundles of inherited genetic material that can only reproduce by hijacking the cells of another life-form. Are viruses alive? The experts disagree.

Life Needs Energy

The sun makes life on Earth possible. Its energy fuels the biosphere. **Photosynthesis** is the process by which plants capture the sun's energy and make it available to other living things. During photosynthesis, plants absorb energy from sunlight and use it to convert water and **carbon dioxide** into food.

Quick Fact!

Only 1 to 2 percent of all the solar energy reaching Earth is captured by plants. The rest is reflected from leaves and/or absorbed by other surfaces and becomes heat.

Photosynthesis happens inside plant cells thanks to a green pigment called chlorophyll. Chlorophyll is one of the most important substances on Earth. It captures the light energy that powers photosynthesis.

Plants are the only land **organisms** that can make food, a kind of chemical energy. Land animals must get the chemical energy they need to survive by eating plants—or animals that eat plants. As herbivores eat plants and carnivores eat other animals, energy moves up the "food chain."

How does energy from the sun move up the food chain to feed a wolf? Grass absorbs the sun's energy and makes food. A deer eats the grass. Then a wolf eats the deer. In this food chain, energy moves from a plant, to an herbivore, to a carnivore.

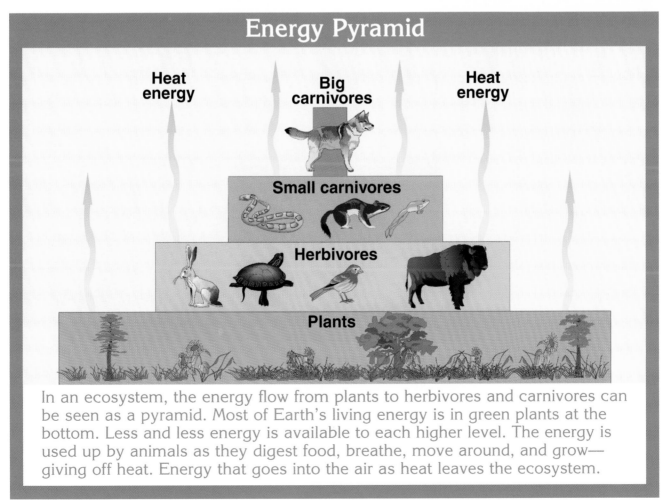

Energy Pyramid

Heat energy

Big carnivores

Heat energy

Small carnivores

Herbivores

Plants

In an ecosystem, the energy flow from plants to herbivores and carnivores can be seen as a pyramid. Most of Earth's living energy is in green plants at the bottom. Less and less energy is available to each higher level. The energy is used up by animals as they digest food, breathe, move around, and grow— giving off heat. Energy that goes into the air as heat leaves the ecosystem.

Life Needs Rock

Rock from Earth's top layer, or crust, makes up the land where most air-breathing animals live. Rock also breaks down into one of the biosphere's key parts—soil. Most plants can't grow without the nutrients found in soil. Without nutrients in soil such as nitrogen and phosphorus, plants can't make food or produce oxygen.

Life not only needs soil, it helps create it. Plant roots help break down, or weather, large rocks into soil over time. Animal wastes and dead plants and animals add organic matter, or humus, to soil.

Some rocks are made up of what was once living. For example, coal is the remains of buried plants, compressed and hardened over millions of years. Limestone is formed from the skeletons and shells of ancient sea life.

Animals need rock, too. They rely on soil-dependent plants for food. They also need certain chemicals that come from rock. For example, animals with bones need phosphorus. You get the nutrient phosphorus from food like cow's milk. The milk cow got its needed phosphorus from the grass it ate. The grass got its phosphorus from rock that broke down into soil. All living things are made up of chemicals like phosphorus that cycle through ecosystems again and again.

The largest living land animal, the elephant, can eat as much as 496 pounds of plant matter in one day.

Quick Fact!

One kilogram (2.2 pounds) of soil can contain 500 billion bacteria, 1 billion fungi, and more than 500 million **invertebrates**, such as worms, mites, and insects.

The soil is home to billions of organisms. Ants alone account for about 15 percent of all land animals by weight.

Investigate Dirt

You need one-quart samples of soil from two different places, 2 lamps with incandescent bulbs, 2 colanders with large holes, 2 large bowls, paper towels, water, a spoon, newspaper, a hand lens, an invertebrate field guide, tape, and a notebook and pencil.

1 Place a paper towel in the bottom of each large bowl and wet it. Set a colander inside each bowl and empty a soil sample into each. Tape a label onto each bowl identifying where the soil came from.

2 Scoop out a spoonful of one of the soils onto some newspaper. *Observe* the soil's characteristics. What color is it? Is it dry or moist? Is its texture sandy, silty, or like clay? Does it have a lot of humus in it, or is it mostly broken-down rock? *Record* your observations.

Humus-rich soil

Clay soil

Sandy soil

Silty soil

3 Repeat step 2 with the other soil sample. *Predict* which soil type has more insects, worms, and other invertebrates living in it.

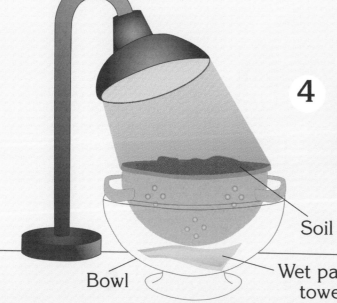

4 Set a lamp about 15 centimeters over each soil-filled colander, as shown. Leave the lamps on at least 24 hours. Keep the paper towel moist and stir the soil occasionally.

Soil

Bowl

Wet paper towel

Earthworm

5 Pick up one of the colanders and *observe* what has crawled out into the bowl. Use the hand lens to observe the smaller organisms, such as mites. Look through the soil in the colander to observe the larger beetles or millipedes. Draw each kind of invertebrate you find and identify it. Then *record* how many of each were in the soil.

Beetle

6 Repeat step 5 with the other soil sample. Compare your findings. Which soil sample contained more animals? Which had more types of animals?

Millipede

Mite

7 Was your prediction correct? What can you *conclude* about the connection between soil types and soil life?

Life Needs Water

Earth's vast oceans make it a blue-colored planet. Water is the most plentiful substance on Earth's surface, covering more than two-thirds, or 71 percent of it.

Water is essential to every organism. The chemical nutrients that fuel and build all life must be dissolved in water for plants and animals to use them. In fact, plants and animals are themselves made up mostly of water. Your body is about 70 percent water.

Quick Fact!

The **hydrosphere** is all the water, ice, and water vapor found in Earth's oceans, lakes, streams, glaciers, ground, and air — some 336 million cubic miles in all.

The world's waters are home to many species of plants, fish, mammals, and invertebrates such as this northern lobster.

Earth's water is always moving between land and air. Rain falls from clouds and fills up lakes and streams, or is soaked up by the soil and plants. The sun heats up lakes and puddles and evaporates them into water vapor in the air. Water is also evaporated out of plant leaves through **transpiration.** All that water vapor in the air eventually forms clouds through **condensation.** When water vapor inside clouds forms ice crystals or water droplets that are heavy enough, they fall as rain, sleet, or snow.

Snow, rain, sleet, and hail are all types of precipitation. Dew and frost are not precipitation. What are they?

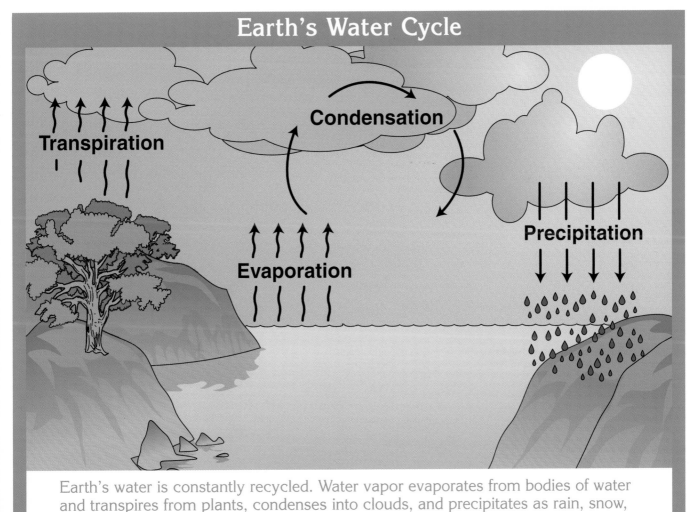

Earth's Water Cycle

Transpiration

Condensation

Evaporation

Precipitation

Earth's water is constantly recycled. Water vapor evaporates from bodies of water and transpires from plants, condenses into clouds, and precipitates as rain, snow, sleet, or hail.

Life Needs Air

Air is the mixture of gases that surrounds Earth—it is its atmosphere. All plants and animals need air to survive. Plants need the carbon dioxide in air for photosynthesis and animals need oxygen to breathe. You can survive for days without water and weeks without food, but only minutes without air.

The atmosphere does more for life than provide air. Life couldn't exist at all without the atmosphere's upper layer of ozone. Ozone is a gas that blocks out deadly doses of the sun's **ultraviolet** rays. The ozone layer protects life on Earth.

Earth's atmosphere extends out about 200 miles before reaching space. But life can survive only in the first couple of miles.

Fish and other aquatic life may not breathe air, but they still need oxygen to live. Their gills and sometimes skin absorb oxygen dissolved in water.

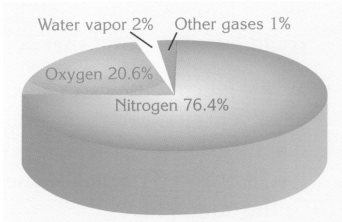

Water vapor 2%   Other gases 1%

Oxygen 20.6%

Nitrogen 76.4%

Earth's air is mostly nitrogen and oxygen. The amount of water vapor varies around the globe, depending on the weather and climate. Other gases include carbon dioxide, argon, and gases found in trace amounts.

Scientists became worried in the late 1970s when they noticed holes developing in Earth's protective ozone layer. They discovered that human-produced chemicals used in refrigerators, air conditioners, and spray cans were causing the destruction of ozone. Now that these chemicals have been banned, scientists believe the ozone layer will repair itself, given time.

Hole in the ozone layer

The atmosphere also affects Earth's temperature. Water vapor and carbon dioxide in the air help warm our world. These **greenhouse gases** trap some of the sun's energy that would otherwise escape into space. Without this greenhouse effect, Earth would be nearly 60°F cooler. Fewer forms of life could live in a world that cold.

While greenhouse gases keep Earth nicely warm, too high a concentration of them would make Earth too hot. Scientists believe that the present global warming trend has likely been caused by the increase of carbon dioxide in the atmosphere.

In the past two hundred years, people have burned more and more coal, gas, and petroleum. These carbon dioxide–spewing fossil fuels have increased the amount of greenhouse gases in the atmosphere.

SCIENTISTS AT WORK:
Predicting Pollution's Effects

Biosphere 2 is a model of living Earth sealed inside a building. Guanghui Lin is Biosphere 2's rain forest expert. "The rain forest is a very important system on Earth," he explains. Today the rain forest stores one-sixth of the carbon dioxide humans put into the air. But as carbon dioxide levels rise, will the rain forest be able to keep up? To find out, Guanghui changed the amount of carbon dioxide inside Biosphere 2's rain forest to what it will likely be by 2050.

What happened? The rain forest couldn't soak up that much carbon dioxide. Like a bucket of water, once it's full, more water can't be added. The bucket just overflows.

Biosphere 2 in Arizona is the world's biggest living laboratory. Inside, scientists can study how Earth's **biomes** work. A biome, such as the rain forest or desert, is a large community of plants and animals that live in a particular climate.

Guanghui Lin controls the environment inside Biosphere 2's rain forest—the rain, the temperature, and the gases in the air—to find how much carbon dioxide this biome can store.

Preserving the Biosphere

Once every ten years, world leaders meet to try to solve the planet's environmental problems. In 2002, for the first time, young delegates attended this meeting—the Earth Summit —and addressed a group of 22,000 presidents, prime ministers, and other people from around the world. They spoke as representatives of your generation.

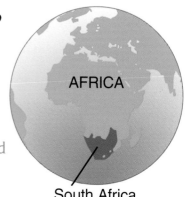

AFRICA

South Africa

The 2002 Earth Summit was held in Johannesburg, South Africa.

Analiz Vergara, a 14-year-old from Ecuador, and 11-year-old Justin Friesen of Canada were two of the official delegates at Earth Summit 2002.

Just as the adult delegates did, Justin and Analiz spoke at meetings and press conferences, and even presented challenges to the General Assembly. For example, they challenged world governments to provide all people with clean drinking water, and they challenged all people to reduce, reuse, and recycle.

In speeches and interviews, Justin and Analiz told world leaders to listen to children's ideas about the environment. They said leaders must take action on their promises to make the world a better place for future generations.

Some delegates visited a community nearby. They met local volunteers who had turned a mountain of trash into a "Mountain of Hope." The volunteers built a traditional-style place where people can sit and tell stories. As a reminder of how the place had once been, some of the trash was used to build this "Hope Tree."

You can work to preserve the biosphere too. Justin says, "I love it when I see people at cleanups, not even knowing who other people are. We just work together to keep our air fresh, our water pure, and our ground clean…. You can help, too…. Do something for the environment today."

Glossary

atmosphere (AT-muh-sfir) mixture of gases that surrounds Earth or other large bodies in space

biome (BYE-ohm) a large community of plants and animals that live in a particular climate such as the rain forest or desert

biosphere (BYE-uh-sfir) the part of Earth, including the air, water, and land, where all organisms live

carbon dioxide (KAR-bun dye-AHK-syd) a colorless, odorless gas used by plants during photosynthesis and breathed out by animals

condensation (kahn-den-SAY-shun) the change of water vapor to a liquid

ecosystem (EE-koh-sis-tum) a community of organisms that interact with each other and the environment they live in

greenhouse gases (GREEN-hows GA-suz) gases in the atmosphere that trap energy from the sun, warming the planet

hydrosphere (HYE-druh-sfir) all of Earth's water, including oceans, lakes and rivers, groundwater, ice, and water vapor in the atmosphere

invertebrate (in-VUR-tuh-brut) an animal without a backbone, such as a worm or insect

organism (OR-guh-nih-zum) a living thing, such as a plant, animal, fungus, or bacterium

photosynthesis (FOH-toh-SIN-thuh-sus) the process by which plants and other organisms that have chlorophyll make food from carbon dioxide and water, using sunlight energy

transpiration (trans-puh-RAY-shun) the flow of water through a plant, into the leaves, and out into the air

ultraviolet (UL-truh-VYE-lut) beyond violet on the color spectrum. Invisible to us, ultraviolet, or UV, rays from the sun can be harmful in large doses.